The Twelve Months

Snowy, Flowy, Blowy,
Showery, Flowery, Bowery,
Hoppy, Croppy, Droppy,
Breezy, Sneezy, Freezy.

Anon.

A Year Full of Poems

Wishes for the Months

I wish you, in January, the swirl of blown snow—
A green January makes a full churchyard;

Thrushes singing through the February rain; in March
The clarion winds, the daffodils;

April, capricious as an adolescent girl,
With cuckoo-song, and cuckoo-flowers;

May with a dog rose, June with a musk rose; July
Multi-foliate, with all the flowers of summer;

August—a bench in the shade, and a cool tankard;
September golden among his sheaves;

In October, apples; in grave November
Offerings for the beloved dead;

And, in December, a midwinter stillness,
Promise of new life, incarnation.

John Heath-Stubbs

A Year Full of Poems

Michael Harrison
and Christopher Stuart-Clark

Oxford University Press

Oxford New York Toronto

Contents

January Jumps About

January jumps about
in the frying pan
trying to heat
his frozen feet
like a Canadian.

February scuttles under
any dish's lid
and she thinks she's dry because she's
thoroughly well hid
but it still rains all month long
and it always did.

March sits in the bath tub
with the taps turned on.
Hot and cold, cold or not,
Has the Winter gone?
In like a lion, out like a lamb
March on, march on, march on.

April slips about
sometimes indoors
and sometimes out
sometimes sheltering from a little
shower of bright rain
in an empty milk bottle
then dashing out again.

May, she hides nowhere,
nowhere at all,
Proud as a peacock
walking by a wall.
The Maytime O the Maytime,
full of leaf and flower.
The Maytime O the Maytime
is loveliest of all.

June discards his shirt and
trousers by the stream
and takes the first dip of the year
into a jug of cream.
June is the gay time
of every girl and boy
who run about and sing and shout
in pardonable joy.

July by the sea
sits dabbling with sand
letting it run out of
her rather lazy hand,
and sometimes she sadly
thinks: 'As I sit here
ah, more than half the year is gone,
the evanescent year.'

August by an emperor
was given his great name.
It is gold and purple
like a Hall of Fame.
(I have known it rather cold
and wettish, all the same.)

September lies in shadows
of the fading summer
hearing, in the distance,
the silver horns of winter
and not very far off
the coming autumn drummer.

October, October
apples on the tree,
the Partridge in the Wood and
the big winds at sea,
the mud beginning in the lane
the berries bright and red
and the big tree wildly
tossing its old head.

November, when the fires
love to burn, and leaves
flit about and fill the air
where the old tree grieves.
November, November
its name is like a star
glittering on many things that were
but few things that are.

Twelfth and last December.
A few weeks away
we hear the silver bells
of the stag and the sleigh
flying from the tundras
far far away
bringing to us all the gift
of our Christmas Day.

George Barker

Winter Wise

Walk fast in snow, in frost walk slow,
And still as you go tread on your toe;
When frost and snow are both together,
Sit by the fire, and spare shoe leather.

Traditional

January

Safe

Come, stir the fire
The lamps unlit
Leave, while we sit
Close to the glow.
And fire and shadow flit
About the room, and fight
For love of it.

Cold winds blow
Whirling in the drear
Night outside; the blaze
Uncoils its tentacles, and here
We in a dream-daze
With the lamps unlit
Safe in firelight sit.

James Walker

Small Birds

Quiet morning, the new year's first day
Written in words of frost upon the almanac:
Dark birds now flutter down, peck away
In circles round each littered barn and stack.
They gather there in busy, greedy throngs,
Eyes full of corn, not caring that it snows;
Sharp hunger puts an edge on all their songs.
The weather turns them into dominoes,
Changes the shape of every tree and shed
As one by one they feed, until the night
Begins to cancel out each toy-like, bobbing head,
And all their world is either black or white.

Leonard Clark

Awakening to Snow

Blades of light slide under my eyelids
and prise them open to discover
softness.
This is whisper day, muffled in deep down
of eiderdown snow, day of
those other echoes and shadows on frosted window-panes
that pass by furtively, wondering.
Today is hushed blue day of nothing, of
empty footprints where feet were,
of absence.
Today adrift from all the gongs of time,
suspended on the feather of your
silent white breath.

Sylvia Kantaris

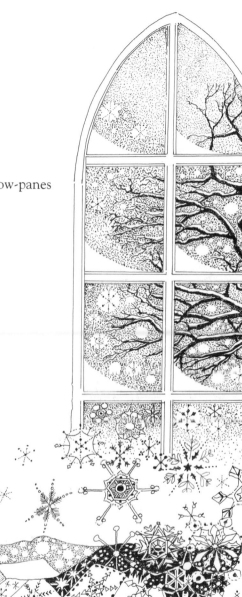

Snowing

Snowing. Snowing. Snowing.
Woolly petals tossed down
From a tremendous tree in the sky
By a giant hand, the hand
That switches on lightning
And tips down cloudbursts.
I like to think of it that way.

Quiet. Quiet. Quiet.
No noise of traffic in the street.
In the classroom only Miss Nil's voice
Dictating and the rustle of paper.
I am holding my breath in wonder.
I want to cry out 'Look! Look!'
Miss Nil has paused between sentences
And is looking out of the window.
But I suppose she is wondering whether
She'll have to abandon her car and walk home.

Snowing. Snowing. Snowing.
I wish I could go out and taste it.
Feel it nestling against my cheek.
And trickling through my fingers.
The message has come round we are to go home now
Because the buses may stop running.
So the snow has given us a whole hour of freedom.
I pick up fistfuls.
Squeeze them hard and hurl them.

But hurry, the bus is coming
And I want to get home early to look at the garden:
At the holly tree in its polar bear coat;
The cherries with white arms upstretched,
Naked of leaves; the scratchy claw marks
Of birds, and blobs of big pawed dogs.
And I want to make footprints of my own
Where the snow is a blank page for scribbling.
Tea time already. Still the snow comes down.
Migrating moths, millions and millions
Dizzying down out of the darkening sky.

Mother draws the curtains.
Why couldn't they stay open?
Now I can't watch the secretive birds
Descending, the stealthy army invading.
What does the roof look like
Covered with slabs of cream?
How high are the heaps on window ledges?
Tomorrow the snow may have begun to melt away.
Why didn't I look more
While there was still time?

Olive Dove

He Who Owns the Whistle Rules the World

january wind and the sun
playing truant again.
Rain beginning to scratch
its fingernails across
the blackboard sky

in the playground
kids divebomb, corner
at Silverstone or execute
traitors. Armed
with my Acme Thunderer
I step outside,
take a deep breath
and bring the world
to a standstill

Roger McGough

The Frozen Man

Out at the edge of town
where black trees

crack their fingers
in the icy wind

and hedges freeze
on their shadows

and the breath of cattle
still as boulders

hangs in rags
under the rolling moon,

a man is walking
alone:

on the coal-black road
his cold

feet
ring

and
ring.

Here in a snug house
at the heart of town

the fire is burning
red and yellow and gold:

you can hear the warmth
like a sleeping cat

breathe softly
in every room.

When the frozen man
comes to the door,

let him in,
let him in,
let him in.

Kit Wright

The Snow

The snow, in bitter cold,
   Fell all the night;
And we awoke to see
   The garden white.

And still the silvery flakes
   Go whirling by,
White feathers fluttering
   From a grey sky.

Beyond the gate, soft feet
   In silence go,
Beyond the frosted pane
   White shines the snow.

F. Ann Elliott

On a Night of Snow

Cat, if you go outdoors you must walk in the snow.
You will come back with little white shoes on your feet,
Little white slippers of snow that have heels of sleet.
Stay by the fire, my Cat. Lie still, do not go.
See how the flames are leaping and hissing low,
I will bring you a saucer of milk like a marguerite,
So white and so smooth, so spherical and so sweet—
Stay with me, Cat. Outdoors the wild winds blow.

Outdoors the wild winds blow, Mistress, and dark is the night.
Strange voices cry in the trees, intoning strange lore;
And more than cats move, lit by our eyes' green light,
On silent feet where the meadow grasses hang hoar—
Mistress, there are portents abroad of magic and might,
And things that are yet to be done. Open the door!

Elizabeth Coatsworth

Explorer

Two o'clock:
Let out of the back door of the house, our cat
Is practising the snow.

The layer of white makes a small, straight, crumbling cliff
Where we open the back door inwards. The cat
Sniffs it with suspicion, learns you can just about
Pat the flaking snow with a careful dab. Then,
A little bolder, he dints it with one whole foot
—and withdraws it, curls it as if slightly lame,

And looks down at it, oddly. The snow is
Different from anything else, not like
A rug, or a stretch of lino, or an armchair to claw upon
And be told to *Get off!*

The snow is peculiar, but not forbidden. The cat
Is welcome to go out in the snow. Does
The snow welcome the cat?
He thinks, looks, tries again.

Three paces out of the door, his white feet find
You sink a little way all the time, it is slow and cold, but it
Doesn't particularly hurt. Perhaps you can even enjoy it, as something new.
So he walks on, precisely, on the tips of very cautious paws . . .

Half-past three, the cat stretched warm indoors.
From the bedroom window we can see his explorations
—From door to fence, from fence to gate, from gate to wall to tree, and back,
Are long pattered tracks and trade-routes of round paw-marks
Which fresh snow is quietly filling.

Alan Brownjohn

When Skies are Low and Days are Dark

When skies are low
and days are dark,
and frost bites
like a hungry shark,
when mufflers muffle
ears and nose,
and puffy sparrows
huddle close—
how nice to know
that February
is something purely
temporary.

N. M. Bodecker

February

February

Splish splosh, February-fill-the-dike,
Sleet in the wind, mud underfoot.
What hint, you ask, of spring? But trust
The honest mistle-thrush, who shouts his song
And builds his nest—a less accomplished singer
Than is the clear-voiced mavis, but he is brave and true.

And trust the aconite and crocus, bright
As wicks of thread which now are lighted up
For ceremonials of Candlemas.

John Heath-Stubbs

Winter Morning

On cold winter mornings
When my breath makes me think
I'm a kettle,
Dad and me wrap up warm
In scarves and balaclavas,
Then we fill a paper bag
With bread and go and feed the ducks
In our local park.
The lake is usually quite frozen
So the ducks can't swim,
They skim across the ice instead,
Chasing the bits of bread
That we throw,
But when they try to peck the crumbs
The pieces slip and slide away.
Poor ducks!
They sometimes chase that bread
For ages and ages,
It makes me hungry just watching them,
So when Dad isn't looking
I pop some bread in my mouth and have a quick chew.
The ducks don't seem to mind,
At least they've never said anything
To me if they do.

Frank Flynn

Cold February

Winter is boring now,
His news is cold,
His hoary anecdotes
Too often told.

For his bad manners we
Made youth the excuse once;
Now in his latter days
He's just a nuisance.

Another season stands
Ready to relieve him;
He's only to be gone
And we'd forgive him,

Even think kindly of him
Finding he's left
A snowdrop morning
As a parting gift.

Hal Summers

Mild February

It is the catkin summer,
The secret season stolen
Under the eye of February,
Whose guarded bags of pollen

The creeping mice of sunbeams
Have nibbled at and holed
Without the iceman noticing:
Out runs the dusty gold.

At any moment—'Hey, what's this?'—
He'll find out what he's lost
And lock his treasuries again
And clang his gates of frost.

Hal Summers

29

Windows

When you look before you go
Outside in the rain or snow,
It looks colder, it looks wetter
Through the window. It is better
When you're outside in it.

When you're out and it's still light
Even though it's almost night
And your mother at the door
Calls you in, there is no more
Daylight in the window
When you're inside looking out.

Russell Hoban

February Twilight

I stood beside a hill
  Smooth with new-laid snow,
A single star looked out
  From the cold evening glow.

There was no other creature
  That saw what I could see—
I stood and watched the evening star
  As long as it watched me.

Sara Teasdale

Ice

The North Wind sighed;
And in a trice
What was water
Now is ice.

What sweet rippling
Water was
Now bewitched is
Into glass:

White and brittle
Where is seen
The prisoned milfoil's
Tender green;

Clear and ringing
With sun aglow,
Where the boys sliding
And skating go.

Now furred's each stick
And stalk and blade
With crystals out of
Dewdrops made.

Worms and ants
Flies, snails and bees
Keep close house guard,
Lest they freeze;

O, with how sad
And solemn an eye
Each fish stares up
Into the sky

In dread lest his
Wide watery home
At night shall solid
Ice become.

Walter de la Mare

Scarcely Spring

Nothing is real. The world has lost its edges;
The sky, uncovered, is the one thing clear.
The earth is little more than atmosphere
Where yesterday were rocks and naked ridges.
Nothing is fixed. Tentative rain dislodges
Green upon green or lifts a coral spear
That breaks in blossom, and the hills appear
Too frail to be the stony fruit of ages.

Nothing will keep. Even the heavens waver.
Young larks, whose first thought is to cry aloud,
Have spent their bubble notes. And here or there
A few slow-hearted boys and girls discover
A moon as insubstantial as a cloud
Painted by air on washed and watery air.

Louis Untermeyer

33

Outside

King Winter sat in his Hall one day,
    And he said to himself, said he,
'I must admit I've had some fun,
I've chilled the Earth and cooled the Sun,
    And not a flower or tree
But wishes that my reign were done,
And as long as Time and Tide shall run,
I'll go on making everyone
    As cold as cold can be.'

There came a knock at the outer door:
    'Who's there?' King Winter cried;
'Open your Palace Gate,' said Spring
'For you can reign no more as King,
    Nor longer here abide;
This message from the Sun I bring,
"The trees are green, the birds do sing;
The hills with joy are echoing":
    So pray, Sir—step outside!'

Hugh Chesterman

March

March

A blue day,
a blue jay
and a good beginning.
One crow, melting snow—
spring's winning!

Elizabeth Coatsworth

MARCH ingorders

Winter has been sacked
for negligence

It appears he left
the sun on all day

Roger McGough

Spring Is

Spring is when
the morning sputters like
bacon
   and
      your
         sneakers
           run
              down
                 the
                    stairs
so fast you can hardly keep up with them
and
spring is when
   your scrambled eggs
      jump
         off
           the
              plate
and turn into a million daffodils
trembling in the sunshine.

Bobbi Katz

A Change in the Year

It is the first mild day of March:
  Each minute sweeter than before,
The redbreast sings from the tall larch
  That stands beside our door.

There is a blessing in the air,
  Which seems a sense of joy to yield
To the bare trees, and mountains bare;
  And grass in the green field.

William Wordsworth

Spring

Spring
slips
silent
snowdrops
past Winter's iron gate.

Then daffodils'
golden trumpets
sound:
Victory!

Hugo Majer

Early Spring

Daffodils shiver,
huddle away from the wind,
like people waiting at a bus-stop.

Adrian Henri

The Spring Flower

I put the bulb in the ground
Weeks and weeks ago
Where it could hide
From winter's frost and snow.
     There it lies
       And shuts its eyes.

And when I had forgotten
That I had put it there
It reached its green leaves out
To test the warmth of the air.
     The bulb sits up in bed
       And puts its hands above its head.

Round a purple flower
Its green leaves opened wide
As if it were opening hands
To show a present inside.
     Up it stands
       And stretches out its hands.

Stanley Cook

First Gathering

Child, take your basket down,
Go and find spring,
Earth has not lost her brown,
Nor wind his sting,
But in the morning
The thrush and the blackbird
Sing to the sleeping town,
And to the waking woods
Sing:
     take your basket down,
Go and find spring!

Now where the ground was bare
Only last week,
Now where the flower was rare
And the hedge bleak,
Reach for the catkin
And stoop for the primrose,
Seek, if you want your share
Of the first gathering,
Seek,
     where the ground was bare
Only last week.

Eleanor Farjeon

The Wind

The wind is a wolf
That sniffs at doors
And rattles windows
With his paws.

Hidden in the night,
He rushes round
The locked-up house,
Making angry sounds.

He leaps on the roof
And tries to drive
Away the house
And everything inside.

Tired next morning,
The wind's still there,
Snatching pieces of paper
And ruffling your hair.

He quietens down and in the end
You hardly notice him go
Whispering down the road
To find another place to blow.

Stanley Cook

Boy with Kite

I am master of my kite, and
the wind tugs against me
on blue ropes of air.
Above tasselled trees
my kite glides and swoops,
pink-and-yellow falcon surging loose
from my tight fist.

White string bites
into flesh; my wrist
flexes like a falconer's.

I am dancing with my kite
heel-and-toe to earth,
body braced
against the fleet north-easter laced
with fraying clouds.

Lifted steeple-clear
of church and school and hill
I am master of my world.

Phoebe Hesketh

April Poem

The morning moon is round and gold
Up in the misty milk-blue sky;
The towering poplar sways alone;
The birds are shouting one by one,
It's first of April, everyone!
In April foolery, that moon
Shines like a small and gentle sun.

Gerda Mayer

April

The First of April

The first of April, some do say,
Is set apart for All Fools' Day,
But why the people call it so
Nor I nor they themselves do know.

Anon.

April Fool

As the April fool came over the hill,
The weather spoke aloud;
It poised a sunny weathercock
Against a livid cloud.

Then the rain came down like pilot-fish,
The lightning bit like sharks,
The grass rushed out of doors to see
All over the green parks.

But the cloud sank from memory,
The blue sail filled with a breeze,
He saw the errands of all birds,
The husbandry of bees.

The ploughland lay as fine as dust,
The sower flung the grains,
The birds came tumbling out of the air
And took their pirate gains.

The pigeon clapped his hands, the gull
Turned living in the sun,
The blackthorn was a thousand pounds;
The April fool looked on.

As the April fool went under the hill
The weather spoke aloud;
It backed an inky weathercock
On a white cotton cloud.

Hal Summers

Springburst

(to be read from the bottom)

FLOWER!
the
slowly   slowly
the *petal* curling
the *bud*,
awakening,

Oh, the
up!
straight
I know!
Now
hm.
see.   hm   see.
me   Hm   me
Let   Let
higher . . .
must reach

for the sky—
Now, must reach
I be!
I live!

up
tip
warmth
coolness
water,
food and
life growing,
life, being,
in the dark—

(seed style)
spark
A

John Travers Moore

To Daffodils

Fair daffodils we weep to see
    You haste away so soon:
As yet the early-rising sun
    Has not attain'd his noon.
     Stay, stay,
    Until the hasting day
     Has run
    But to the evensong;
And, having pray'd together, we
    Will go with you along.

We have short time to stay, as you,
    We have as short a spring;
As quick a growth to meet decay,
    As you, or any thing.
     We die,
    As your hours do, and dry
     Away,
    Like to the summer's rain;
Or as the pearls of morning's dew
    Ne'er to be found again.

Robert Herrick

Smells

Through all the frozen winter
My nose has grown most lonely
For lovely, lovely, colored smells
That come in springtime only.

The purple smell of lilacs,
The yellow smell that blows
Across the air of meadows
Where bright forsythia grows.

The tall pink smell of peach trees,
The low white smell of clover,
And everywhere the great green smell
Of grass the whole world over.

Kathryn Worth

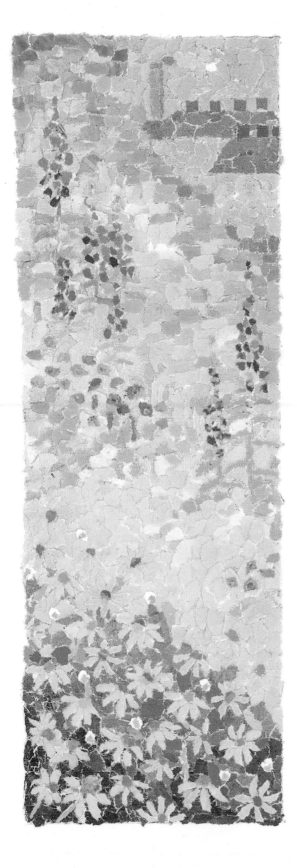

The Flowers in Town

Among the busy streets
In the middle of the town
Is a flowery field
Where houses have been knocked down.

The men with cranes and bulldozers
Left the ground brown and bare
Except for the broken bricks
Scattered everywhere.

The ground was rough and bumpy
And there the old bricks lay
Like a set of building blocks
That hadn't been put away.

But the seeds of flowers
That were looking for a home
Travelled there on the wind
And made the place their own.

Ragweed that seems to be knitted
Out of yellow wool
And poppies like red crêpe paper
Have filled the hollows full.

High in the air, the willow herb
Raises its pointed towers
And daisies and butterfingers
Pattern the grass with flowers.

Where the people used to live
In the houses the men knocked down
Bees and butterflies are busy
In the flowers' new town.

Stanley Cook

Tadpoles

For J.W.

Since my first infant term I remember
their surprise arrivals brought a pleasure
keen and not predicted by the calendar.

One morning you would see glass glint, feel squirms
of fascination rippling into grins.
Our teachers took down old aquariums

from cupboard tops. We rinsed them clean of dust
then filled them with sieved pond-water at first.
After that for topping up the tap sufficed.

One overflowing year Mike Cotton came
with an enamel bucket brimming spawn—
too much for classroom conservation.

Everybody took a handful
in lunchbox, jar or milkbottle.
My plastic bag leaked a trickle

of sticky seepage that ran down
my legs and prompted curious frowns
on faces in the bus back home.

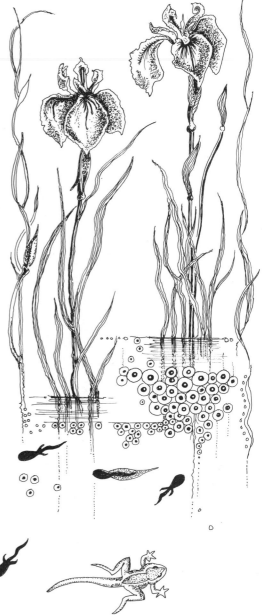

They hatched and grew fat on the windowsill
safe in a flower-vase untouched until
(come round to help a bit) my Aunty Jill

poured them accidentally down the toilet,
then said the family must have something desperate.
My mother never quite forgave her that.

Philip Tupper fed them one by one
to his piranha. Lizzie kept hers in
a bowl until her puppy drank them. When

Jemima's died she pressed them all like leaves
between the pages of a book. Now these
dippings into childhood pools draw symphonies

into a single jar intense and bright
with remembered water, sky and weed and light:
the crotchets quivering their silent music.

Barrie Wade

I am the Rain

I am the rain
I like to play games
like sometimes
      I pretend
I'm going
      to fall
Man that's the time
I don't come at all

Like sometimes
I get these laughing stitches
up my sides
      rushing people in
and out
      with the clothesline
I just love drip
      dropping
down collars
      and spines
Maybe it's a shame
but it's the only way
I get some fame

Grace Nichols

April Rain Song

Let the rain kiss you.
Let the rain beat upon your head with silver liquid drops.
Let the rain sing you a lullaby.

The rain makes still pools on the sidewalk.
The rain makes running pools in the gutter.
The rain plays a little sleep-song on our roof at night—

And I love the rain.

Langston Hughes

Early Spring

Once more the Heavenly Power
    Makes all things new,
And domes the red-plow'd hills
    With loving blue;
The blackbirds have their wills,
    And throstles too.

Opens a door in Heaven;
    From skies of glass
A Jacob's ladder falls
    On greening grass,
And o'er the mountain-walls
    Young angels pass.

Before them fleets the shower,
    And burst the buds,
And shine the level lands,
    And flash the floods;
The stars are from their hands
    Flung thro' the woods,

The woods with living airs
    How softly fann'd,
Light airs from where the deep,
    All down the sand,
Is breathing in his sleep,
    Heard by the land.

For now the Heavenly Power
    Makes all things new,
And thaws the cold, and fills
    The flower with dew;
The blackbirds have their wills,
    The poets too.

Alfred, Lord Tennyson

The Four Sweet Months

First, April, she with mellow showers
Opens the way for early flowers;
Then after her comes smiling May,
In a more sweet and rich array;
Next enters June, and brings us more
Gems than those two that went before:
Then, lastly, July comes, and she
More wealth brings in than all those three.

Robert Herrick

May

Fields of golden rape,
melting
in the intensity
of their own glowing colour,
slowly
butter the hillsides.

Roger McGough

May

May-Time

There is but one May in the year,
    And sometimes May is wet and cold;
There is but one May in the year,
    But before the year grows old.

Yet, though it be the chilliest May
    With least of sun, and most of showers,
Its wind and dew, its night and day,
    Bring up the flowers.

Anon.

As I Went Out

As I went out this May morning I found
A ruby in the grass;
So deep and pure a ray could never come
From broken bits of glass.

Near by, two sapphires lay,
Thrown down at dead of night,
And they too shone with an intense, clear beam,
Two very eyes of light.

A little further off were emeralds
Pale, dark and twice as many,
A topaz here, and as for diamonds,
Well, they were two a penny.

Who was it threw them down,
Bound on what errand or what escapade?
I wonder about him: was he bold and careless
Or thievish and afraid?

Gathered they could not be,
Possessed they might if eyes
And memory can possess, and so they can,
Not otherwise—

Or so I thought, but then came driving by
On his delivery round the sun
Who picked them up like empty milk-bottles;
By midday all were gone.

Hal Summers

Cricketer

Light
as the flight
of a bird on the wing
my feet skim the grass
and my heart seems to sing:
'How green is the wicket.
It's cricket.
It's spring.'

Maybe the swallow
high in the air
knows what I feel
when I bowl fast and follow
the ball's twist and bounce.
Maybe the cat
knows what I feel like, holding my bat
and ready to pounce.
Maybe the tree
so supple and yielding
to the wind's sway
then swinging back, gay,
might know the way
I feel when I'm fielding.

Oh, the bird, the cat and the tree:
they're cricket, they're me.

R. C. Scriven

Sky

Tall and blue
true and open

So open my arms have room
for all the world
for sun and moon
            for birds and stars

Yet how I wish I had the chance
to come drifting down to earth—
            a simple bed sheet
covering some little girl or boy
just for a night
            but I am Sky
            that's why

Grace Nichols

The Dream of the Cabbage Caterpillars

There was no magic spell:
    all of us, sleeping,
dreamt the same dream—a dream
    that's ours for the keeping.

In sunbeam or dripping rain,
    sister by brother,
we once roamed with glee
    the leaves that our mother

laid us and left us on,
    browsing our fill
of green cabbage, fresh cabbage,
    thick cabbage, until

in the hammocks we hung
    from the garden wall
came sleep, and the dream
    that changed us all—

we had left our soft bodies,
    the munching, the crawling,
to skim through the clear air
    like white petals falling!

Just so, so we woke—
    so to skip high as towers,
and dip now to sweet fuel
    from trembling bright flowers.

Libby Houston

The Caterpillar

Brown and furry
Caterpillar in a hurry;
Take your walk
To the shady leaf or stalk.

May no toad spy you,
May the little birds pass by you;
Spin and die,
To live again a butterfly.

Christina Rossetti

The Butterfly

There is no story behind it.
It is split like a second.
It hinges around itself.

It has no future.
It is pinned down to no past.
It's a pun on the present.

It's a little yellow butterfly.
It has taken these wretched hills
under its wings.

Just a pinch of yellow,
it opens before it closes
and closes before it o

where is it

Jejuri Arun Kolatkar

May

Now children may
   Go out of doors,
Without their coats,
   To candy stores.

The apple branches
   And the pear
May float their blossoms
   Through the air,

And Daddy may
   Get out his hoe
To plant tomatoes
   In a row,

And, afterwards,
   May lazily
Look at some baseball
   On TV.

John Updike

The Rain

I don't care what you say
I like
the rain,
I like it chucked like nails
against my win-
dow pane.

I don't care what you say
I like
the soak
of drizzle when it drifts
about the hills
like smoke.

I don't care what you say
I like
the flood
that makes our road the Nile
and leaves our lawn
all mud.

I don't care what you say
I like
it, so,
with hat and coat and yellow wellies, here
I go

on holiday to find
that fam-
ous plain
in Spain. I don't care what
you say, my friend,
I like
the rain.

Richard Edwards

Have You Heard the Sun Singing?

Have you ever heard the sun in the sky
Man have you heard it?
Have you heard it break the black of night
Man have you heard it?
Have you heard it shouting its song, have you heard
It scorch up the air like a phoenix bird,
Have you heard the sun singing?

John Smith

June

Summer

Rushes in a watery place,
   And reeds in a hollow;
A soaring skylark in the sky,
   A darting swallow;
And where pale blossom used to hang
   Ripe fruit to follow.

Christina Rossetti

What is the Sun?

the Sun is an orange dinghy
    sailing across a calm sea

it is a gold coin
    dropped down a drain in Heaven

the Sun is a yellow beach ball
    kicked high into the summer sky

it is a red thumb-print
    on a sheet of pale blue paper

the Sun is a milk bottle's gold top
    floating in a puddle

Wes Magee

The School-Boy

I love to rise in a summer morn,
When the birds sing on every tree;
The distant huntsman winds his horn,
And the sky-lark sings with me.
O! what sweet company.

But to go to school in a summer morn,
O! it drives all joy away;
Under a cruel eye outworn,
The little ones spend the day,
In sighing and dismay.

Ah! then at times I drooping sit,
And spend many an anxious hour,
Nor in my book can I take delight,
Nor sit in learning's bower,
Worn thro' with the dreary shower.

How can the bird that is born for joy,
Sit in a cage and sing?
How can a child when fears annoy,
But droop his tender wing,
And forget his youthful spring?

O! father and mother, if buds are nip'd,
And blossoms blown away,
And if the tender plants are strip'd
Of their joy in the springing day,
By sorrow and care's dismay,

How shall the summer arise in joy
Or the summer fruits appear?
Or how shall we gather what griefs destroy
Or bless the mellowing year,
When the blasts of winter appear?

William Blake

Timeless

There is no clock in the forest
but a dandelion to blow,
an owl that hunts
when the light has gone,
a mouse that sleeps
till night has come,
lost in the moss below.

There is no clock in the forest,
only the cuckoo's song
and the thin white
of the early dawn,
the pale damp-bright
of a waking June,
the bluebell light
of a day half-born
when the stars have gone.

There is no clock in the forest.

Judith Nicholls

Adlestrop

Yes, I remember Adlestrop—
The name, because one afternoon
Of heat the express-train drew up there
Unwontedly. It was late June.

The steam hissed. Someone cleared his throat.
No one left and no one came
On the bare platform. What I saw
Was Adlestrop—only the name

And willows, willow-herb, and grass,
And meadowsweet, and haycocks dry,
No whit less still and lonely fair
Than the high cloudlets in the sky.

And for that minute a blackbird sang
Close by, and round him, mistier,
Farther and farther, all the birds
Of Oxfordshire and Gloucestershire.

Edward Thomas

The Rainy Summer

There's much afoot in heaven and earth this year;
　　The winds hunt up the sun, hunt up the moon,
Trouble the dubious dawn, hasten the drear
　　Height of a threatening noon.

No breath of boughs, no breath of leaves, of fronds,
　　May linger or grow warm; the trees are loud;
The forest, rooted, tosses in her bonds,
　　And strains against the cloud.

No scents may pause within the garden-fold;
　　The rifled flowers are cold as ocean-shells;
Bees, humming in the storm, carry their cold
　　Wild honey to cold cells.

Alice Meynell

To Walk in Warm Rain

To walk in warm rain
　　And get wetter and wetter!
To do it again—
To walk in warm rain
　　Till you drip like a drain.
To walk in warm rain
　　And get wetter and wetter.

David McCord

What Could Be Lovelier than to Hear

What could be lovelier than to hear
The summer rain
Cutting across the heat, as scythes
Cut across grain?
Falling upon the steaming roof
With sweet uproar,
Tapping and rapping wildly
At the door?

No, do not lift the latch,
But through the pane
We'll stand and watch the circus pageant
Of the rain,
And see the lightning, like a tiger,
Striped and dread,
And hear the thunder cross the sky
With elephant tread.

Elizabeth Coatsworth

July

The Crab has got the sun in its claws.
Julius Caesar makes the laws;
His legions march for a Roman cause—
　　Hurrah for Julius Caesar!

The mountains shake at the legions' tread,
Like the sound of thunder overhead;
There's an old bald codger at their head—
　　His name is Julius Caesar.

In Caesar's month the July sun
Knows summer's course will soon be run,
But the sun has always work to be done—
　　'Like me,' said Julius Caesar.

John Heath-Stubbs

July

Rain in Summer

How beautiful is the rain!
After the dust and heat,
In the broad and fiery street,
In the narrow lane,
How beautiful is the rain!
How it clatters along the roofs,
Like the tramp of hoofs!

How it gushes and struggles out
From the throat of the overflowing spout!
Across the window pane
It pours and pours;
And swift and wide,
With a muddy tide,
Like a river down the gutter roars
The rain, the welcome rain!

Henry Wadsworth Longfellow

Summer School

Summer sun bakes
mirage meadows
on our playground.

Its ripples gild
the asphalt pools
with barley ears.

Sparrow flickers
spread the fresh-cut,
clean smell of grass.

The teacher drones.

Flies drowse
at the edge of hearing.

Gnat clouds shimmer
in light behind
our closing lids.

The lesson swings
its scythe-sweeps
at nettles overgrown.

Its wordclouds blur
and separate,
then drift away

beyond the leafmould
in our minds'
vacated nests.

My name is called.

Quick!
the harvest mice

are in full flight
before the flash
of all-consuming blades.

Barrie Wade

July

Loud is the Summer's busy song,
The smallest breeze can find a tongue,
While insects of each tiny size
Grow teasing with their melodies,
Till noon burns with its blistering breath
Around, and day lies still as death.

The busy noise of man and brute
Is on a sudden lost and mute;
Even the brook that leaps along,
Seems weary of its bubbling song.
And, so soft its waters creep,
Tired silence sinks in sounder sleep;

The cricket on its bank is dumb;
The very flies forget to hum;
And, save the wagon rocking round,
The landscape sleeps without a sound.
The breeze is stopped, the lazy bough
Hath not a leaf that danceth now;

The taller grass upon the hill,
And spider's threads, are standing still;
The feathers, dropped from moorhen's wing
Which to the water's surface cling,
Are steadfast, and as heavy seem
As stones beneath them in the stream;

Noon swoons beneath the heat it made,
And flowers e'en within the shade;
Until the sun slopes in the west,
Like weary traveller, glad to rest
On pillowed clouds of many hues.
Then Nature's voice its joy renews,

And checkered field and grassy plain
Hum with their summer songs again,
A requiem to the day's decline,
Whose setting sunbeams coolly shine
As welcome to day's feeble powers
As falling dews to thirsty flowers.

John Clare

Once the Wind

Once the wind
said to the sea
I am sad
      And the sea said
Why
      And the wind said
Because I
am not blue like the sky
or like you

     So the sea said what's
so sad about that
         Lots
of things are blue
or red or other colours too
     but nothing
neither sea nor sky
can blow so strong
or sing so long as you

    And the sea looked sad
     So the wind said
Why

Shake Keane

Flotilla

Curved clouds are sailing, like yachts with spinnakers,
Across the blue, calm reach of sky:
A fleet of white, scudding before the wind.

Lying on the slow turf, fathoms down,
We watch them move serenely through the heads
Of oak and elm, out to the open sea.

Who knows what gales, what disasters will meet them there,
Beyond the range of our eyes? But still they come,
Passing with billowing sails in endless procession.

Clive Sansom

What the Wind Said

'Far away is where I've come from,' said the wind
'Guess what I've brought you.'
   'What?' I asked.
'Shadows dancing on a brown road by an old
Stone fence,' the wind said. 'Do you like that?'
   'Yes,' I said. 'What else?'
'Daisies nodding, and the drone of one small airplane
In a sleepy sky,' the wind continued.
   'I like the airplane, and the daisies too,' I said.
   'What else!'
'That's not enough?' the wind complained.
   'No,' I said. 'I want the song that you were singing.
   Give me that.'
'That's mine,' the wind said. 'Find your own.' And left.

Russell Hoban

Summer

Winter is cold-hearted,
    Spring is yea and nay,
Autumn is a weathercock
        Blown every way.
    Summer days for me
When every leaf is on its tree;

    When Robin's not a beggar,
    And Jenny Wren's a bride,
And larks hang singing, singing, singing
    Over the wheat-fields wide,
    And anchored lilies ride,
    And the pendulum spider
    Swings from side to side;

And blue-black beetles transact business,
    And gnats fly in a host,
And furry caterpillars hasten
    That no time be lost,
    And moths grow fat and thrive,
    And ladybirds arrive.

    Before green apples blush,
    Before green nuts embrown,
    Why one day in the country
    Is worth a month in town;
    Is worth a day and a year
Of the dusty, musty, lag-last fashion
    That days drone elsewhere.

Christina Rossetti

The Harvest Moon

The flame-red moon, the harvest moon,
Rolls along the hills, gently bouncing,
A vast balloon,
Till it takes off, and sinks upward
To lie in the bottom of the sky, like a gold doubloon.

The harvest moon has come,
Booming softly through heaven, like a bassoon.
And earth replies all night, like a deep drum.

So people can't sleep,
So they go out where elms and oak trees keep
A kneeling vigil, in a religious hush.
The harvest moon has come!

And all the moonlit cows and all the sheep
Stare up at her petrified, while she swells
Filling heaven, as if red hot, and sailing
Closer and closer like the end of the world.

Till the gold fields of stiff wheat
Cry 'We are ripe, reap us!' and the rivers
Sweat from the melting hills.

Ted Hughes

In the Green Shade

I wade by the edge of the sparkling stream
Plunged waist-down in a sea of green
The still spread of dock and wild-rhubarb leaves
I can push to one side and peek between

At another world, where the half-light is green
And the stiff stalks and the arching leaves
Make cloisters I almost could crawl between
On the damp dark earth that slopes to the stream.

What would it be like to live under the leaves
The sunlight never filters between?
To slide like a snail, silent, leaving a stream
Of moisture behind me, white on the green?

To be one of the slug-herd, and creep in between
My shiny brethren, moo down to the stream,
And drink in the dark, or be one of the green
Bat-insects that hang upside-down from the leaves?

I shiver, and stand. The once-sparkling stream
Runs grey under trees no longer so green
For the sun has gone that shone through their leaves
And a chill wind has drawn black clouds in between.

Brian Lee

Pods Pop and Grin

Strong strong sun, in that look
you have, lands ripen
fruits, trees, people.

Lands love the flame of your gaze.
Lands hide some warmth
of sun-eye for darkness.

All for you pods pop and grin.
Bananas hurry up and grow.
Coconut becomes water and oil.

Palm trees try to fly to you
but just dance everywhere.
Silk leaves of bamboo rustle wild.

And when rain finished falling
winds shake diamonds from branches
that again feel your eye.

Strong strong sun, in you
lands keep ripening
fruits, trees, people.

Birds go on tuning up
and don't care at all—
more blood berries are coming.

Your look strokes up all
summertime. We hear streams running.
You come back every day.

James Berry

Summer

When it's hot
I take my shoes off,
I take my shirt off,
I take my pants off,
I take my underwear off,
I take my whole body off,
and throw it
in the river.

Frank Asch

August

August

The sprinkler twirls.
   The summer wanes.
The pavement wears
   Popsicle stains.

The playground grass
   Is worn to dust.
The weary swings
   Creak, creak with rust.

The trees are bored
   With being green.
Some people leave
   The local scene

And go to seaside
   Bungalows
And take off nearly
   All their clothes.

John Updike

Harvest

I saw the farmer plough the field,
And row on row
The furrows grow.
I saw the farmer plough the field,
And hungry furrows grow.

I saw the farmer sow the wheat,
The golden grain,
In sun and rain.
I saw the farmer sow the wheat,
In shining sun and rain.

I saw at first a silvery sheen,
Then line on line
Of living green.
I saw at first a silvery sheen,
Then lines of living green.

The living green then turned to gold,
In thirty—fifty—
Hundred fold.
The living green then turned to gold
In mercies manifold.

M. M. Hutchinson

August

The city dwellers all complain
When August comes and brings no rain.
The pavements burn upon their feet;
Temper and temperature compete.
They mop their brows, they slow their pace,
And wish they were some other place.

But farmers do not mind the heat;
They know it ripens corn and wheat.
They love to see the sun rise red,
Remembering what their fathers said:
'An August month that's dry and warm
Will never do the harvest harm.'

Michael Lewis

Until I Saw the Sea

Until I saw the sea
I did not know
that wind
could wrinkle water so.

I never knew
that sun
could splinter a whole sea of blue.

Nor
did I know before
a sea breathes in and out
upon a shore.

Lilian Moore

Spray

It is a wonder foam is so beautiful.
A wave bursts in anger on a rock, broken up
in wild white sibilant spray
and falls back, drawing in its breath with rage,
with frustration how beautiful!

D. H. Lawrence

Going Barefoot

With shoes on,
I can only feel
how hard or soft
the rock or sand is
where I walk
or stand.

Barefoot,
I can feel
how warm mud
moulds my soles—
or how cold
pebbles
knead them
like worn knuckles.

Curling my toes,
I can drop
an anchor
to the sea floor—
hold fast
to the shore
when the tide
tows.

Judith Thurman

Let's Hear it for the Limpet

If there's one animal that isn't a wimp, it
Is the limpet.

Let me provide an explanation
For my admiration.

To start with, it's got two thousand tiny teeth
Beneath

Its comical conical-hat-shaped, greeny-grey shell:
A tongue as well

That rasps the delicate seaweed through its front door:
What's more—

And this is what gives me the greatest surprise—
Two bright eyes

Indoors at the end of long tentacles poking out, which
Twitch.

But its funniest feature by far is its foot
That's put

Straight down to clamp fast to the rock.
(Gulls knock,

You see, at the shell to try and winkle it off
For scoff.)

Kit Wright

Tell me, Tell me, Sarah Jane

Tell me, tell me, Sarah Jane,
    Tell me, dearest daughter,
Why are you holding in your hand
    A thimbleful of water?
Why do you hold it to your eye
    And gaze both late and soon
From early morning light until
    The rising of the moon?

Mother, I hear the mermaids cry,
    I hear the mermen sing,
And I can see the sailing-ships
    All made of sticks and string.
And I can see the jumping fish,
    The whales that fall and rise
And swim about the waterspout
    That swarms up to the skies.

Tell me, tell me, Sarah Jane,
    Tell your darling mother,
Why do you walk beside the tide
    As though you loved none other?
Why do you listen to a shell
    And watch the billows curl,
And throw away your diamond ring
    And wear instead the pearl?

Mother, I hear the water
    Beneath the headland pinned,
And I can see the sea-gull
    Sliding down the wind.
I taste the salt upon my tongue
    As sweet as sweet can be.

Tell me, my dear, whose voice you hear?

    It is the sea, the sea.

Charles Causley

maggie and milly and molly and may

maggie and milly and molly and may
went down to the beach(to play one day)

and maggie discovered a shell that sang
so sweetly she couldn't remember her troubles,and

milly befriended a stranded star
whose rays five languid fingers were;

and molly was chased by a horrible thing
which raced sideways while blowing bubbles:and

may came home with a smooth round stone
as small as a world and as large as alone.

For whatever we lose(like a you or a me)
it's always ourselves we find in the sea

e e cummings

Tides

It's time to go, but still we sit
Lingering in our summer
Like idle fingers,
Like fingers in the sand.

Or like a tiny snail that moves
Beneath a gravelly pool,
Taking its life to travel,
Taking between the tides.

John Fuller

Little Fish

The tiny fish enjoy themselves
in the sea.
Quick little splinters of life,
their little lives are fun to them
in the sea.

D. H. Lawrence

Thistledown

Silver against blue sky
These ghosts of day float by,
Fitful, irregular,
Each one a silk-haired star,
Till from the wind's aid freed
They settle on their seed.

Not by the famished light
Of a moon-ridden night
But by clear sunny hours
Gaily these ghosts of flowers
With rise and swirl and fall
Dance to their burial.

Andrew Young

August Ends

A nip in the air today, and autumn
Playing hide and seek with summer;
Winter takes a first grip on plant, insect, bird.
Last blackberry flowers fade,
And fruit, moving from green to red,
Dangles foot long purple clusters
Over downy hedgerows, wasps go numb,
Fall drowsy on dropped plums, honey and smoky wax
Perfume the spidered loft, barley shines.
Swifts on curved wings wheel overhead
Printing broad arrows on the leaden sky;
And now I catch the echo of the far north wind
And over the shorn and stubbed land
The dreaded hawk hovers, and a cloud of peewits cry.

Leonard Clark

Glass Falling

The glass is going down. The sun
Is going down. The forecasts say
It will be warm, with frequent showers.
We ramble down the showery hours
And amble up and down the day.
Mary will wear her black galoshes
And splash the puddles on the town;
And soon on fleets of macintoshes
The rain is coming down, the frown
Is coming down of heaven showing
A wet night coming, the glass is going
Down, the sun is going down.

Louis MacNeice

September

And Now

It's never *now* it's always *wait until...*
But this September Sunday, under Falkland Hill,
Cornstalks were stacked in rolls and rolls
And the sun struck gold in these ochre bales:
That moment hangs there still though I pass on,
Chasing my shadow the length of the sun.

Alan Bold

New Boy's View of Rugger

When first I played I nearly died.
    The bitter memory still rankles—
They formed a scrum with *me* inside!
    Some kicked the ball and some my ankles.
I did not like the game at all,
    Yet, after all the harm they'd done me,
Whenever I came near the ball
    They knocked me down and stood upon me.

Rupert Brooke

Are You Ready?

It's
 September
  the
   sixth,
    the
     day
     before
    school,
   we
  go
 back
  tomorrow
   and
    I
     feel
      like
     a
    fool.
   I
  can't
 find
 my
  bag,
   my
    ruler,
     my
    pen.

I
 can
  hardly
  recall
  if
   I'm
    Andy
     or
      Ken!
     I'm
    all
   of
  a
 dither,
  tomorrow's
   a
    haze,
     the
      school
     starts
    in
     hours
    and
   I'm
  in
   a
    daze.

Wes Magee

Summer Goes

Summer goes, summer goes
Like the sand between my toes
When the waves go out.
That's how summer pulls away,
Leaves me standing here today,
Waiting for the school bus.

Summer brought, summer brought
All the frogs that I have caught,
Frogging at the pond,
Hot dogs, flowers, shells and rocks,
Postcards in my postcard box—
Places far away.

Summer took, summer took
All the lessons in my book,
Blew them far away.
I forgot the things I knew—
Arithmetic and spelling too,
Never thought about them.

Summer's gone, summer's gone—
Fall and winter coming on,
Frosty in the morning.
Here's the school bus right on time.
I'm not really sad that I'm
Going back to school.

Russell Hoban

September

The golden-rod is yellow;
    The corn is turning brown;
The trees in apple orchards
    With fruit are bending down.

The gentian's bluest fringes
    Are curling in the sun;
In dusky pods the milkweed
    Its hidden silk has spun.

The sedges flaunt their harvest
    In every meadow nook;
And asters by the brookside
    Make asters in the brook.

From dewy lanes at morning
    The grapes' sweet odors rise;
At noon the roads all flutter
    With golden butterflies.

By all these lovely tokens
    September days are here,
With summer's best of weather,
    And autumn's best of cheer.

Helen Hunt Jackson

Autumn

Now the summer is grown old
the light long summer
                       is grown old.
Leaves change
and the garden is gold
with marigolds and zinnias
tangled and bold
blazing blazing
orange and gold.
             The light long summer
       is grown old.

Charlotte Zolotow

Rain

The lights are all on, though it's just past midday,
There are no more indoor games we can play,
No one can think of anything to say,
It rained all yesterday, it's raining today,
It's grey outside, inside me it's grey.

I stare out of the window, fist under my chin,
The gutter leaks drips on the lid of the dustbin,
When they say 'cheer up', I manage a grin,
I draw a fish on the glass with a sail-sized fin,
It's sodden outside, and it's damp within.

Matches, bubbles and papers pour into the drains,
Clouds smother the sad laments from the trains,
Grandad says it brings on his rheumatic pains,
The moisture's got right inside of my brains,
It's raining outside, inside me it rains.

Brian Lee

Just Another Autumn Day

In Parliament, the Minister
for Mists and Mellow Fruitfulness
announces, that owing to
inflation and rising costs
there will be no Autumn
next year. September, October
and November are to be
cancelled, and the Government
to bring in the nine-month year instead.
Thus will we all live longer.

Emergency measures are to be
introduced to combat outbreaks
of well-being, and feelings
of elation inspired by the season.
Breathtaking sunsets will be
restricted to alternate Fridays
and gentle dusks prohibited.
Fallen leaves will be outlawed
and persons found in possession
of conkers, imprisoned without trial.
Thus will we all work harder.

The announcement caused little reaction.
People either way don't really care
No time have they to stand and stare
Looking for work or slaving away
Just another Autumn day.

Roger McGough

October

October turned my maple's leaves to gold;
    The most are gone now; here and there one lingers.
Soon these will slip from out the twig's weak hold,
    Like coins between a dying miser's fingers.

Thomas Bailey Aldrich

October

October

The year slows down. The swallows go,
Leaving our valley far below
Floating in mist. Nests in the eaves
Are empty, the gutters choked with leaves.
There are berries on the bryony,
The hawthorn and the rowan-tree;
The squirrel now forgets to swing,
The fieldmouse stops his scampering,
Searching in every hole and rut
For beechmast, acorn, hazelnut.
Even the butterflies are slow
In their brown wanderings to and fro . . .
And later, frosts will come, to take
The rings and ripples from the lake
And lend her, as those wrinkles pass,
The smooth transparency of glass.

Clive Sansom

From The Bramble Hedge

It is two o'clock. The pin-head spider
is sure his life is long, as long
as his six legs. The fly
feeding on blackberries is convinced
he is meeting God, eye to eye.

Joan Downar

Fly Away, Fly Away

Fly away, fly away over the sea,
Sun-loving swallow, for summer is done;
Come again, come again, come back to me,
Bringing the summer and bringing the sun.

Christina Rossetti

Crowfield

They are holding
council meetings demonstrations
in the angry air
picket the October sunlight.
Flakes of burnt paper
fall towards the stubblefield.

Adrian Henri

October Tuesday

One crow in a high wind over Chelsea,
black against a rain sky loops and swings,
writes, '*Black against a rain sky*' with its wings.
One leaf, blown yellowing upward over Paultons Square,
writes '*Winter soon, yes, winter*' on the air.

Russell Hoban

An October Wind

An October wind
Ruffles the heavy shade of
The summer maple

Its leaves grow
Lightheaded and
Drop off in a million
Crimson circles

The tree is an
Outline of itself
Good for
Unruffling the wind

Zaro Weil

Rags

The night wind
rips a cloud sheet
into rags,

then rubs, rubs
the October moon
until it shines
like a brass doorknob.

Judith Thurman

The Wild Swans at Coole

The trees are in their autumn beauty,
The woodland paths are dry,
Under the October twilight the water
Mirrors a still sky;
Upon the brimming water among the stones
Are nine-and-fifty swans.

The nineteenth autumn has come upon me
Since I first made my count;
I saw, before I had well finished,
All suddenly mount
And scatter wheeling in great broken rings
Upon their clamorous wings.

I have looked upon those brilliant creatures,
And now my heart is sore.
All's changed since I, hearing at twilight,
The first time on this shore,
The bell-beat of their wings above my head,
Trod with a lighter tread.

Unwearied still, lover by lover,
They paddle in the cold,
Companionable streams or climb the air;
Their hearts have not grown old;
Passion or conquest, wander where they will,
Attend upon them still.

But now they drift on the still water,
Mysterious, beautiful;
Among what rushes will they build,
By what lake's edge or pool
Delight men's eyes when I awake some day
To find they have flown away?

W. B. Yeats

Leaves

Who's killed the leaves?
Me, says the apple, I've killed them all.
Fat as a bomb or a cannonball
I've killed the leaves.

Who sees them drop?
Me, says the pear, they will leave me all bare
So all the people can point and stare.
I see them drop.

Who'll catch their blood?
Me, me, me, says the marrow, the marrow.
I'll get so rotund that they'll need a wheelbarrow.
I'll catch their blood.

Who'll make their shroud?
Me, says the swallow, there's just time enough
Before I must pack all my spools and be off.
I'll make their shroud.

Who'll dig their grave?
Me, says the river, with the power of the clouds
A brown deep grave I'll dig under my floods.
I'll dig their grave.

Who'll be their parson?
Me, says the Crow, for it is well-known
I study the bible right down to the bone.
I'll be their parson.

Who'll be chief mourner?
Me, says the wind, I will cry through the grass
The people will pale and go cold when I pass.
I'll be chief mourner.

Who'll carry the coffin?
Me, says the sunset, the whole world will weep
To see me lower it into the deep.
I'll carry the coffin.

Who'll sing a psalm?
Me, says the tractor, with my gear grinding glottle
I'll plough up the stubble and sing through my throttle.
I'll sing the psalm.

Who'll toll the bell?
Me, says the robin, my song in October
Will tell the still gardens the leaves are over.
I'll toll the bell.

Ted Hughes

Gathering Leaves

In autumn the falling leaves
Run races on the paths,
Tumble head over heels
And catch against the tufts of grass.

I gather them in a heap
With a stiff brush and a rake,
Though they are light as feathers
And do their best to escape.

Then I splash right into the heap
And the leaves wash over me
With a long swishing sound
Like a wave of the sea.

Stanley Cook

Song

Fall, leaves, fall; die, flowers, away;
Lengthen night and shorten day;
Every leaf speaks bliss to me
Fluttering from the autumn tree.
I shall smile when wreaths of snow
Blossom where the rose should grow;
I shall sing when night's decay
Ushers in a drearier day.

Emily Brontë

Leaves in the Yard

Leaves have the lightest footfall
Of all who come in the yard.
They play rounders, they play tig,
They play no-holds-barred.

Late, when people are all asleep
Still they scamper and weave.
They play robbers, they play cops,
They play Adam-and-Eve.

Tap, tap, on the pavement,
Flit, flit, in the air:
The sentry-going bat wonders what they're at,
The blank back-windows stare.

When they rest, the wind rests;
When they go, he goes too;
They play tiptoe, they play mouse,
He shouts *hoo*.

Summer, they fidgeted on trees,
Then autumn called 'Enough!'
They play leapfrog, they play fights,
They play blind-man's-buff.

Ragged, swept in corners,
Fallen beyond recall,
Ragged and old, soon to be mould,—
But light of heart wins all.

Hal Summers

Conkers

Out of sight they spend whole summers
growing spiky in the leaf corners.
We never hear them drop:
their swell and fall
is secret as imagination.
In split shells they lie,

nuggets for polishing.
Damp from casings briefly clings
like mist across the sunrise.
They burnish in our hands,
send bubbles through the blood,
make minds molten with joy.

They are poems, varied
and irresistible,
each containing
its own new germination.
Arranging them on strings
will thread our Autumn through with fire.

Barrie Wade

A Day in Autumn

It will not always be like this,
The air windless, a few last
Leaves adding their decoration
To the trees' shoulders, braiding the cuffs
Of the boughs with gold; a bird preening
In the lawn's mirror. Having looked up
From the day's chores, pause a minute,
Let the mind take its photograph
Of the bright scene, something to wear
Against the heart in the long cold.

R. S. Thomas

A Hallowe'en Pumpkin

They chose me from my brother: 'That's the
Nicest one,' they said,
And they carved me out a face and put a
Candle in my head;

And they set me on the doorstep. Oh, the
Night was dark and wild;
But when they lit the candle, then I
Smiled!

Dorothy Aldis

From My Window

The waters fall in rectangles
Cities sprout umbrellas
Clear water paints a blurred picture
Tears slide from afternoon branches
Dirt breathes a deep sigh of mud
It's raining today

Zaro Weil

November

November

The shepherds almost wonder where they dwell,
And the old dog for his right journey stares:
The path leads somewhere, but they cannot tell,
And neighbour meets with neighbour unawares.
The maiden passes close beside her cow,
And wanders on, and thinks her far away;
The ploughman goes unseen behind his plough
And seems to lose his horses half the day.
The lazy mist creeps on in journey slow;
The maidens shout and wonder where they go;
So dull and dark are the November days.
The lazy mist high up the evening curled,
And now the morn quite hides in smoke and haze;
The place we occupy seems all the world.

John Clare

Rainy Nights

I like the town on rainy nights
When everything is wet—
When all the town has magic lights
And streets of shining jet!

When all the rain about the town
Is like a looking-glass,
And all the lights are upside down
Below me as I pass.

In all the pools are velvet skies,
And down the dazzling street
A fairy city gleams and lies
In beauty at my feet.

Irene Thompson

City Rain

After the storm
all night before
the world looked like
an upturned mop

wrung out into streets
half-dirty, half-clean,
tasting of rain
in bedraggled trees

and smelling of dog
with its shaky fur
and cold

lick.

Kit Wright

November

The leaves are fading and falling,
　　The winds are rough and wild,
The birds have ceased their calling,
　　But let me tell you, my child,

Though day by day, as it closes,
　　Doth darker and colder grow,
The roots of the bright red roses
　　Will keep alive in the snow.

And when the Winter is over,
　　The boughs will get new leaves,
The quail come back to the clover,
　　And the swallow back to the eaves.

The robin will wear on his bosom
　　A vest that is bright and new,
And the loveliest way-side blossom
　　Will shine with the sun and dew.

The leaves today are whirling,
　　The brooks are dry and dumb,
But let me tell you, my darling,
　　The Spring will be sure to come.

There must be rough, cold weather,
　　And winds and rains so wild;
Not all good things together
　　Come to us here, my child.

So, when some dear joy loses
　　Its beauteous summer glow,
Think how the roots of the roses
　　Are kept alive in the snow.

Alice Cary

The Old Field

The old field is sad
Now the children have gone home.
They have played with him all afternoon,
Kicking the ball to him, and him
Kicking it back.

But now it is growing cold and dark.
He thinks of their warm breath, and their
Feet like little hot water bottles.
A bit rough, some of them, but still . . .

And now, he thinks, there's not even a dog
To tickle me.
The gates are locked.
The birds don't like this nasty sneaking wind;
And nor does he.

D. J. Enright

Ploughing

The tractor-driver ploughs his road as straight as a Roman's,
Changing rough stubble into smooth, shining earth,
Turning it over in waves that break and fall motionless.

A white pennant of gulls follows, shrilling and screaming.
The angry birds drop to the furrow, snatch at their food,
And lift again like torn papers blown in the wind.

Clive Sansom

The Sun and Fog Contested

The Sun and Fog contested
The Government of Day—
The Sun took down his yellow whip
And drove the Fog away.

Emily Dickinson

Fog in November

Fog in November, trees have no heads,
Streams only sound, walls suddenly stop
Half-way up hills, the ghost of a man spreads
Dung on dead fields for next year's crop.
I cannot see my hand before my face,
My body does not seem to be my own,
The world becomes a far-off, foreign place,
People are strangers, houses silent, unknown.

Leonard Clark

No!

No sun—no moon!
No morn—no noon—
No dawn—no dusk—no proper time of day—
No sky—no earthly view—
No distance looking blue—
No road—no street—no 't'other side the way'—
No top to any steeple—
No recognitions of familiar people—
No courtesies for showing 'em—
No knowing 'em!
No travelling at all—no locomotion—
No inkling of the way—no notion—
'No go'—by land or ocean—
No mail—no post—
No news from any foreign coast—
No park—no ring—no afternoon gentility—
No company—no nobility—
No warmth, no cheerfulness, no healthful ease,
No comfortable feel in any member—
No shade, no shine, no butterflies, no bees,
No fruits, no flowers, no leaves, no birds.
November!

Thomas Hood

Sleet

The first snow was sleet. It swished heavily
Out of a cloud black enough to hold snow.
It was fine in the wind, but couldn't bear to touch
Anything solid. It died a pauper's death.

Now snow—it grins like a maniac in the moon.
It puts a glove on your face. It stops gaps.
It catches your eye and your breath. It settles down
Ponderously crushing trees with its airy ounces.

But today it was sleet, dissolving spiders on cheekbones,
Being melting spit on the glass, smudging the mind
That humped itself by the fire, turning away
From the ill wind, the sky filthily weeping.

Norman MacCaig

December

Winter

Winter crept
through the whispering wood,
hushing fir and oak;
crushed each leaf and froze each web—
but never a word he spoke.

Winter prowled
by the shivering sea,
lifting sand and stone;
nipped each limpet silently—
and then moved on.

Winter raced
down the frozen stream,
catching at his breath;
on his lips were icicles,
at his back was death.

Judith Nicholls

A Peculiar Christmas

Snow? Absolutely not.
In fact, the weather's quite hot.
At night you can watch this new
Star without catching the 'flu.

Presents? Well, only three.
But then there happen to be
Only three guests. No bells,
No robins, no fir-trees, no smells

—I mean of roast turkey and such:
There are whiffs in the air (a bit much!)
Of beer from the near public house,
And of dirty old shepherds, and cows.

The family party's rather
Small—baby, mother and father—
Uncles, aunts, cousins dispersed.
Well, this Christmas *is* only the first.

Roy Fuller

The Robin in December

When the leaves have fallen
And the days begin to shorten;
When the dark night draws its curtains
At tea-time on the sun;
When the summer flowers have gone
And we put our warm coats on—
The robin comes back to the garden.

All the rest of the year
We knew he was somewhere near
And we saw him from time to time
On the wall or the lawn or flying by,
But he never came before
Right up to the kitchen door.

Is he waiting to be drawn
To go on a Christmas card?
Is he bringing a touch of red
Now most of the roses are dead?
No: we guess why he comes
And put out seed for him and crumbs.

Stanley Cook

Christmas Daybreak

Before the paling of the stars,
    Before the winter morn,
Before the earliest cockcrow,
    Jesus Christ was born:
Born in a stable,
    Cradled in a manger,
In the world His hands had made,
    Born a stranger.

Priest and king lay fast asleep
    In Jerusalem,
Young and old lay fast asleep
    In crowded Bethlehem:
Saint and angel, ox and ass,
    Kept a watch together,
Before the Christmas daybreak
    In the winter weather.

Jesus on His Mother's breast
 In the stable cold,
Spotless Lamb of God was He,
 Shepherd of the fold.
Let us kneel with Mary Maid,
 With Joseph bent and hoary,
With saint and angel, ox and ass,
 To hail the King of Glory.

Christina Rossetti

Advice to a Child

Set your fir-tree
In a pot;
Needles green
Is all it's got.
Shut the door
And go away,
And so to sleep
Till Christmas Day.
In the morning
Seek your tree,
And you shall see
What you shall see.

Hang your stocking
By the fire,
Empty of
Your heart's desire;
Up the chimney
Say your say,
And so to sleep
Till Christmas Day.
In the morning
Draw the blind,
And you shall find
What you shall find.

Eleanor Farjeon

The Christmas Tree

They chopped her down in some far wood
A week ago,
Shook from her dark green spikes her load
Of gathered snow,
And brought her home at last, to be
Our Christmas show.

A week she shone, sprinkled with lamps
And fairy frost;
Now, with her boughs all stripped, her lights
And spangles lost,
Out in the garden there, leaning
On a broken post,

She sighs gently . . . Can it be
She longs to go
Back to that far-off wood, where green
And wild things grow?
Back to her dark green sisters, standing
In wind and snow?

John Walsh

Before Christmas

The year tips, the sun
slips towards the sky's edge, and
dark bites at the day.

Shopping after dark:
hands clutching carrier-bags
stuffed with surprises.

The pillar-box's
smiling mouth swallows our cards,
cheered by the greetings.

Christmas cards snow through
the letter-box. Open them
and brightness thaws out.

Lying awake I
hear clattering hooves: reindeer
landing on the roof.

John Corben

After Christmas

Darkness begins a
retreat: the cold light flows back
over the dead land.

Put the tree out now:
hang nuts on its branches—see feathered
decorations come.

Take down the Christmas
cards: arrowheads in the dust
point to spring cleaning.

Pull down the paper
chains: the room grows tall, the floor
deep in coloured snow.

Cold bites deep: warm your
mind at Christmas memories
and look for snowdrops.

John Corben

The Old Year

The Old Year's gone away
   To nothingness and night:
We cannot find him all the day
   Nor hear him in the night:
He left no footstep, mark or place
   In either shade or sun:
The last year he'd a neighbour's face,
   In this he's known by none.

All nothing everywhere:
   Mists we on mornings see
Have more of substance when they're here
   And more of form than he.
He was a friend by every fire,
   In every cot and hall—
A guest to every heart's desire,
   And now he's nought at all.

Old papers thrown away,
   Old garments cast aside,
The talk of yesterday,
   All things identified;
But times once torn away
   No voices can recall:
The eve of New Year's Day
   Left the Old Year lost to all.

John Clare

Change

The summer
still hangs
heavy and sweet
with sunlight
as it did last year.

The autumn
still comes
showering gold and crimson
as it did last year.

The winter
still stings
clean and cold and white
as it did last year.

The spring
still comes
like a whisper in the dark night.

It is only I
who have changed.

Charlotte Zolotow

Index of Authors

Index of Titles and First Lines

First lines are in italics

The Artists

The illustrations are by:

Camilla Charnock   pp 12–13, 24–25, 37, 104–
    105
Jacky Corner   endpapers, pp 22–23, 47,
    110–111
John Craig   pp 53, 125
Robina Green   pp 38–39, 50–51, 87, 112, 128
Zoë Hancox   pp 74–75, 90–91, 107
Tudor Humphries   pp 30–31, 78–79, 126–127
Valerie McBride   pp 14–15, 26–27, 34–35,
    42–43, 54–55, 62–63, 70–71, 82–83, 94–95,
    102–103, 114–115, 122–123, 134–135
Alan Marks   pp 20, 41, 84–85, 96
Jackie Morris   pp 18–19, 58–59, 81, 130–131
Deryk Thomas   pp 33, 69, 120, 121, 133
Tracy Thompson   pp 29, 66–67, 98–99, 116,
    118–119
Sarah van Niekerk   pp 108, 109
Mary van Riemsdyk   pp 17, 48–49, 61, 76, 92,
    93

The jacket illustration is by Valerie McBride

Acknowledgements

The editors and publisher are grateful for permission to reproduce the following copyright material:

Frank Asch: 'Summer' reprinted from *Country Pie*, © 1979 by Frank Asch, by permission of Greenwillow Books, William Morrow & Co. **George Barker:** 'January Jumps About' from *To Aylsham Fair*. Reprinted by permission of Faber & Faber Ltd. **James Berry:** 'Pods Pop and Grin' from *When I Dance* (Hamish Hamilton 1988), © 1988 James Berry. Reprinted by permission of the publisher. **N. M. Bodecker:** 'When Skies are Low and Days are Dark' from *Snowman Sniffles*. © 1983 N. M. Bodecker. Reprinted by permission of Faber & Faber Ltd. and Margaret K. McElderry Books, an imprint of Macmillan Publishing Co. **Alan Bold:** 'And Now'. Reprinted by permission of the author. **Alan Brownjohn:** 'Explorer'. Reprinted by permission of the author. **Charles Causley:** 'Tell me, tell me, Sarah Jane' from *Collected Poems* (Macmillan). Reprinted by permission of David Higham Associates Ltd. **Hugh Chesterman:** 'Outside' from *Book of a Thousand Poems* (Evans Bros.). © Mrs Hugh Chesterman. **Leonard Clark:** 'Small Birds' and 'Fog in November' from *Six of the Best* (Puffin). Reprinted by permission of Robert Clark, Literary Executor. 'August Ends' from *Good Company*. Reprinted by permission of Dobson Books Ltd. **Elizabeth Coatsworth:** 'On a Night of Snow', 'March' and 'What Could be Lovelier Than to Hear'. © Catherine Beston Barnes. **Stanley Cook:** 'The Spring Flowers', 'The Wind', 'The Robin in December', 'Gathering Leaves' and 'The Flowers in Town' from *The Squirrel in Town*. Reprinted by permission of Blackie and Son Ltd. **John Corben:** 'Before Christmas' and 'After Christmas'. Reprinted by permission of the author. **e e cummings:** 'maggie and millie and molly and may' from *Complete Poems 1913–1962*, © 1923, 1925, 1931, 1935, 1938, 1939, 1940, 1944, 1945, 1946, 1947, 1948, 1949, 1950, 1951, 1952, 1953, 1954, 1955, 1956, 1957, 1958, 1959, 1960, 1961, 1962 by the Trustees for the e e cummings Trust. © 1961, 1963, 1968 by Marion Morehouse Cummings. Reprinted by permission of Grafton Books, a division of HarperCollins Publishers Ltd., and the Liveright Publishing Corp. **Walter de la Mare:** 'Ice'. Reprinted by permission of The Literary Trustees of Walter de la Mare and The Society of Authors as their representative. **Olive Dove:** 'Snowing', first published in *Drumming in the Sky* (BBC). Reprinted by permission of the author. **Joan Downar:** 'From the Bramble Hedge'. Reprinted from *The Old Noise of Truth* (1989) by permission of Peterloo Poets. **Richard Edwards:** 'The Rain'. Reprinted by permission of the author. **D. J. Enright:** 'The Old Field' from *Rhyme Times Rhyme* (Chatto & Windus). Reprinted by

permission of Watson Little Ltd., Author's agents. **Eleanor Farjeon:** 'First Gathering' and 'Advice to a Child' from *The Children's Bells* (OUP). Reprinted by permission of David Higham Associates Ltd. **Frank Flynn:** 'Winter Morning' from *The Candy Floss Tree*. Reprinted by permission of the author. **John Fuller:** 'Tides'. Reprinted by permission of the author. **Roy Fuller:** 'A Peculiar Christmas'. Reprinted from *The World Through the Window* by permission of Blackie & Son Ltd. **John Heath-Stubbs:** 'Wishes for the Months', 'February' and 'July' from *Naming the Beasts* (Carcanet). Reprinted by permission of David Higham Associates Ltd. **Adrian Henri:** 'Early Spring' from *The Phantom Lollipop Lady* (Methuen) and 'Crowfield' from *Rhinestone Rhino*. Reprinted by permission of Rogers Coleridge & White Ltd. **Phoebe Hesketh:** 'Boy with Kite' reprinted from *Netting the Sun: New and Collected Poems* by permission of Enitharmon Press. **Russell Hoban:** 'Windows', 'October Tuesday', 'What the Wind Said' and 'Summer Goes' from *Egg Thoughts and Other Frances Songs* (Faber). Reprinted by permission of David Higham Associates Ltd. **Libby Houston:** 'The Dream of the Cabbage Caterpillars'. Reprinted by permission of the author. **Langston Hughes:** 'April Rain Song' from *The Dream Keeper and Other Poems*. © 1932 by Alfred A. Knopf Inc., and renewed 1960 by Langston Hughes. Reprinted by permission of the publisher and Harold Ober Associates. **Ted Hughes:** 'The Harvest Moon' and 'Leaves' from *Season Songs*. Text © 1975 by Ted Hughes. Reprinted by permission of Faber & Faber Ltd. **Sylvia Kantaris:** 'Awakening to Snow' reprinted by permission of Bloodaxe Books Ltd. from *Dirty Washing: New & Selected Poems* by Sylvia Kantaris (Bloodaxe Books 1989). **Bobbi Katz:** 'Spring Is', © 1979, used with permission of Bobbi Katz, New York, copyright holder. **Shake Keane:** 'Once the Wind' from *Poetry Jump-Up* (Blackie & Son Ltd). **Brian Lee:** 'In the Green Shade' and 'Rain' from *Late Home* (Kestrel 1976), © Brian Lee, 1976. Reprinted by permission of Penguin Books Ltd. **Michael Lewis:** 'August'. **Norman MacCaig:** 'Sleet'. Reprinted from *Collected Poems* (Chatto & Windus) by permission of Random Century Ltd. **Louis MacNeice:** 'Glass Falling' from *The Collected Poems of Louis MacNeice*. Reprinted by permission of Faber & Faber Ltd. **David McCord:** 'To Walk in Warm Rain' from *Speak Up*, © 1979, 1980 by David McCord. Reprinted by permission of Little, Brown and Co. **Roger McGough:** 'He who owns the whistle' from *In the Glassroom* (Jonathan Cape), and 'Just Another Autumn Day' from *Holiday in Death Row* (Jonathan Cape). Reprinted by permission of Random Century Ltd., and the Peters Fraser & Dunlop Group. 'MARCH ingorders' and 'May' from *Nailing the Shadow* (Viking Kestrel). Reprinted by permission of the Peters Fraser & Dunlop Group Ltd. **Wes Magee:** 'Are You Ready?' and 'What is the Sun?' reprinted from *The Witch's Brew and other poems* (Cambridge Univ. Press 1989), by permission of the author and publisher. **Hugo Majer:** 'Spring'. Reprinted by permission of the author. **Gerda Mayer:** 'April Poem'. © 1989 Gerda Mayer. Reprinted by permission of the author. **John Travers Moore:** 'Springburst' from *There's Motion Everywhere,* © 1970 by John Travers Moore and published by the Houghton Mifflin Co. Used by permission of the author. **Lilian Moore:** 'Until I Saw the Sea' from *I Feel the Same Way.* © 1967 by Lilian Moore. Reprinted by permission of Marian Reiner for the author. **Judith Nicholls:** 'Timeless', © 1990 Judith Nicholls, from *Dragonsfire* by Judith Nicholls, published by Faber, 1990. Reprinted by permission of the author. 'Winter' from *Midnight Forest.* Reprinted by permission of Faber & Faber Ltd. **Grace Nichols:** 'I am the Rain' and 'Sky' from *Come on into my Tropical Garden.* © Grace Nichols 1988. Reprinted by permission of Curtis Brown Group Ltd. on behalf of Grace Nichols. **Clive Sansom:** 'Flotilla', 'October' and 'Ploughing' from *An English Year* (Chatto & Windus). Reprinted by permission of David Higham Associates Ltd. **R. C. Scriven:** 'Cricketer' from *The Livelong Day.* © R. C. Scriven. Reprinted by permission of the author. **John Smith:** 'Have you heard the sun singing?' reprinted in *Poems to Paddle In* (Hutchinson). **Hal Summers:** 'Cold February', 'Mild February', 'April Fool', 'Leaves in the Yard' and 'As I Went Out', © Hal Summers 1978, from *Tomorrow is my Love* by Hal Summers (1978). Reprinted by permission of Oxford University Press. **Sara Teasdale:** 'February Twilight' from *Collected Poems,* © 1926 by Macmillan Publishing Co., renewed 1954 by Mamie T. Wheless. Reprinted by permission of Macmillan Publishing Co. **R. S. Thomas:** 'A Day in Autumn'. Reprinted from *Song at the Year's Turning,* © R. S. Thomas, 53 Gloucester Road, Kew, UK. Used with permission. **Judith Thurman:** 'Going Barefoot' and 'Rags' from *Flashlight and other poems* (Atheneum). © 1976 by Judith Thurman. **Louis Untermeyer:** 'Scarcely Spring' from *Burning Bush,* © 1928 by Harcourt Brace Jovanovich, Inc. and renewed 1956 by Louis Untermeyer. Reprinted by permission of the publisher. **John Updike:** 'May' and 'August' from *A Child's Calendar.* © 1965 by John Updike and Nancy Burkert. Reprinted by permission of Alfred A. Knopf, Inc. **Barrie Wade:** 'Tadpoles', 'Conkers' and 'Summer School', © Barrie Wade 1989, reprinted from *Conkers* (1989) by permission of Oxford University Press. **James Walker:** 'Safe' reprinted in *Tapestry* (Edward Arnold). Copyright Fortune Press. **John Walsh:** 'The Christmas Tree' from *Poets in Hand* (Puffin). Reprinted by permission of Mrs A. M. Walsh. **Zaro Weil:** 'An October Wind' and

'From My Window', © Zaro Weil 1989, reproduced from *Mud, Moon and Me* by kind permission of Orchard Books, 96 Leonard Street, London EC2A 4RH. **Kathryn Worth:** 'Smells' from *The Walker Book of Poetry for Children.* © Kathryn Worth. **Kit Wright:** 'The Frozen Man' from *Rabbiting On.* Reprinted by permission of Collins Publishers. 'Let's hear it for the Limpet' and 'City Rain' from *Cat Among the Pigeons* (Viking Kestrel 1987) © Kit Wright, 1987. Reprinted by permission of Penguin Books Ltd. **W. B. Yeats:** 'The Wild Swans at Coole', from *Collected Poems,* © 1919 by Macmillan Publishing Co., renewed 1947 by Bertha Georgie Yeats. Reprinted by permission of Macmillan Publishing Co. **Andrew Young:** 'Thistledown'. Reprinted by permission of Alison Young, Literary Executor. **Charlotte Zolotow:** 'Autumn' and 'Change' from *River Winding,* text © 1970 by Charlotte Zolotow. Reprinted by permission of HarperCollins Publishers, New York. Published in the UK by World's Work/Heinemann.

While we have tried to secure reprint permission prior to publication this has not always proved possible. Full source details are given where known and if notified the publisher will be pleased to amend or correct the credits at the earliest opportunity.

Oxford University Press, Walton Street, Oxford OX2 6DP

Oxford New York Toronto
Delhi Bombay Calcutta Madras Karachi
Petaling Jaya Singapore Hong Kong Tokyo
Nairobi Dar es Salaam Cape Town
Melbourne Auckland

and associated companies in
Berlin Ibadan

Oxford is a trade mark of Oxford University Press

This selection and arrangement ©
Michael Harrison and Christopher Stuart-Clark 1991

First published 1991
First published by Oxford in the United States 1991

Library of Congress Catalog Card Number: 91 052812
A CIP catalogue record for this book is available
from the British Library

ISBN 0 19 276097 1

Typeset by Pentacor PLC, High Wycombe, Bucks
Printed in Hong Kong